U.S DEPARTMENT OF COMMERCE

Bureau of Industry and Security

A Guide to Export Licensing Requirements

CONTENTS

A FREQUENTLY ASKED QUESTIONS GUIDE TO EXPORT LICENSING REQUIREMENTS FOR COMMERCIAL ITEMS

This brochure is designed to give people who are new to exporting, and, in particular, new to export controls, a general understanding of Department of Commerce regulations and how to use them. However, nothing provided here can substitute for consulting the Export Administration Regulations.

Does My Shipment Require an Export License?

Maybe. A U.S. export license requirement from the Department of Commerce can be triggered by several important factors specific to your transaction: the actual item (commodity, software or technology) being exported, where it is going, who is going to use it, and what they will be using it for. If any of these factors change in your transaction, the license requirements may change.

What types of items does the Department of Commerce regulate?

The Bureau of Industry and Security (BIS) implements and enforces the Export Administration Regulations (EAR). The EAR regulate the export and reexport of most commercial items. We often refer to the items that BIS regulates as "dual-use" items because they have both commercial and military or proliferation applications, but purely commercial items without an obvious military use also are subject to the EAR.

Does the Department of Commerce regulate all exports?

The Department of Commerce does not regulate all goods, services, and technologies. Other U.S. Government agencies have export control responsibilities for regulating more specialized exports. For example, if you are shipping military goods, your item may be subject to the licensing jurisdiction of the Directorate of Defense Trade Controls at the Department of State. The Treasury Department's Office of Foreign Assets Controls (OFAC) administers and enforces economic and trade sanctions against targeted foreign countries, terrorism sponsoring organizations, and international narcotics traffickers. The BIS website identifies resource links for various U.S. Government agencies with export control responsibilities. Go to http://www.bis.doc.gov/about/reslinks.htm for a listing.

Is there a list of restricted countries to which I can not export?

Restrictions vary from country to country and from item to item. The most restricted destinations are the embargoed countries and those countries designated as supporting terrorist activities, including Cuba, Iran, North Korea, Sudan, and Syria.

So how do I know if my shipment needs an export license?

The first step is knowing your item's Export Control Classification Number (ECCN). ECCN entries are found on the Commerce Control List (CCL) and identify reasons for control which indicate licensing requirements to certain destinations. Other reasons an export license may be required for your shipment relate to concerns about the parties to the transaction and the end-use of the item.

Do all items have an ECCN?

Many commercial goods are not on the Commerce Control List and do not have an ECCN. These goods are designated as EAR99. EAR99 items generally consist of low-level technology, consumer goods, etc. and do not require a license in most situations. However, if your proposed export of an EAR99 item is to an embargoed country, to an end-user of concern, or in support of a prohibited end-use, you may be required to obtain an export license.

Where do I find the Commerce Control List?

The Commerce Control List is part of the Export Administration Regulations (EAR) which can be found on the Government Printing Office's EAR database at **Error! Hyperlink reference not valid.**. Scroll down the page to Part 774. Below are the ten categories of the Commerce Control List which can be viewed as separate files in several different formats, including PDF.

So if my item is EAR99, does that mean I don't need a license?

EAR99 items will generally ship under the designation "NLR" which stands for "No License Required". However, if your proposed export of an EAR99 item is to an embargoed country, to an end-user of concern or in support of a prohibited end-use, you may be required to obtain an export license.

What is the difference between EAR99 and NLR?

EAR99 is a classification for an item. It indicates that a particular item is subject to the Export Administration Regulations (EAR), but not specifically described by an Export Control Classification Number (ECCN) on the Commerce Control List

(CCL). While the classification describes the item, the authorization for shipment of that item may change, depending on the circumstances of the transaction.

NLR stands for the "No License Required" designation. NLR may be used for either EAR99 items, or items on the CCL that do not require a license for the destination. However, exports of an EAR99 item to an embargoed country, an end-user of concern or in support of a prohibited end-use may require an export license.

If I determine my item is classified EAR99 and I can ship under NLR, what do I need to do?

You indicate "NLR" as your authorization for export on the Shipper's Export Declaration or Automated Export System record. By signing the SED or submitting via AES (or designating your freight forwarder to do so), you are certifying that your item is eligible for NLR. You should also keep records relating to your NLR determination and the export transaction for five years.

What is a license exception?

A license exception is an authorization that allows you to export or reexport, under stated conditions, items subject to the Export Administration Regulations (EAR) that would otherwise require a license based on the ECCN and destination.

What do you mean by prohibited end-user or end-use?

If you know or have reason to know your item would support a proliferation activity, such as nuclear, chemical/biological, or missile proliferation activities in a country of concern, a license would be required. Part 744 of the Export Administration Regulations spells out the specific regulations related to end-user and end-use controls.

Do the licensing requirements change depending on how I'm sending the item?

No. Export license requirements stay the same regardless of the method of shipment or transmission. This includes technology shipments via the Internet or items carried in a briefcase. [Note: although license requirements may not change, license exception availability may (i.e. BAG or TMP.)]

I'm not a company, I'm just sending something overseas to a friend. Does this mean I don't need to worry about whether or not my item needs a license?

Whether or not something is being given away or sold does not affect the license requirements of your shipment. You still need to determine whether or not an export license is required.

Does it matter if what I'm sending is under $2500 dollars?

No. The value of the shipment does not affect the export license requirements. However, the value of the shipment may affect the requirements for filing a SED or AES Record. If an export license is required, a Shipper's Export Declaration will be required as well, regardless of value.

Is a shipment to an Army Post Office (APO) or Fleet Post Office (FPO) considered an export?

Yes. These shipments are exports because they are being shipped to destinations outside the United States. License requirements are those that are applicable to the destination. Certain license exceptions may also be available.

Can I just have my freight forwarder fill out the license related information on the Shippers Export Declaration or AES record?

Your freight forwarder is generally not in a position to determine your requirements because a technical assessment of your item is required to determine its ECCN and any associated license requirements. You are ultimately responsible for the proper export authorization.

What if my customer asks me to send the item to their freight forwarder here in the U.S.? Do I still have obligations?

Yes. Just because your customer is directing the U.S. forwarder to ship the items does not relieve you of obligations in the export transaction. The responsibility for determining the proper export authorization and obtaining an export license, if necessary, would be your obligation unless your customer provides you something in writing indicating that they are assigning the licensing responsibility to another party in the U.S., such as the forwarder. In either case, you will continue to show as the "Exporter" on the Shipper's Export Declaration. In this case you need to provide specific information to the forwarder in order for them to do what it takes to ensure compliance. You will still be cited on the export declaration as U.S. Principal Party in Interest (USPPI).

I've got some paperwork from past exports made before I started here, can I just use the same information again?

Not necessarily. Export license requirements are transaction specific. If the item, country of destination, end-user or end-use have changed, it could affect the type of authorization for which the export is eligible.

My company has been exporting for years and I've never heard of an ECCN or license requirement. Is this requirement new?

No, this requirement is not new. Perhaps someone in your company already determined that the items you export are not listed in a specific ECCN on the Commerce Control List and do not require export licenses for the destinations to which you ship. Be aware that classifications may change over time. Since the exporter is responsible for determining the proper authorization for shipment, you will want to review the Commerce Control List or submit a classification request to BIS.

If you find there have been exports without proper authorization, you should notify your company's management and consider filing a voluntary disclosure to BIS.

How do I get an ECCN number for my item?

There are three ways to obtain an ECCN number: (1) Contact the manufacturer as they may know the ECCN of their item; (2) Self classify the item; or (3) Submit an official commodity classification request to BIS electronically using SNAP-R.

I know the Schedule B number for my item, will that help in determining the ECCN?

No. The ECCN and Schedule B number do not correlate, although you will need to know both when you make a shipment.

What is the Schedule B Number and how do I get it?

Schedule B commodity codes are 10-digit numeric codes used to identify products for trade statistics purposes. The Bureau of Census's Foreign Trade Division can assist you at 301-763-2238. A Schedule B search engine is available at http://www.census.gov/ foreign-trade/schedules/b/.

I've been told that I can I get an official ECCN determination over the phone from the Department of Commerce. Is this true?

No. The ECCN is based on the technical characteristics of the item and requires a detailed analysis of the item in order for it to be classified. The Department of Commerce cannot provide you with an ECCN over the telephone. However, we can assist you in understanding how to do a self-classification of the item using on-line resources and your technical understanding of the item. We can also explain the process of submitting an official request for a classification which typically takes 4-6 weeks. The Office of Exporter Services has counselors available from our Washington DC headquarters at (202) 482-4811 and our Western Regional Office in California at (949) 660-0144 or (408) 291-4212.

I don't have time to wait for a response to my classification request. Are there any alternatives to this?

If you are not the manufacturer of the item you are exporting, the item may have already been classified by the manufacturer if they themselves export. One of the quickest ways to determine the ECCN of your item is to check with the manufacturer. Many companies list ECCNs on their website.

I've found an official classification done by the Department of Commerce back in 1996. Can I still use this?

The Commerce Control List and ECCNs change often - sometimes items are added to the CCL and other times item specifications are changed or removed. In addition, your item may have different technical characteristics than what was classified previously. It's important to stay up-to-date with changes to the Commerce Control List which may affect your item's classification.

Once I've determined my ECCN, what do I do?

Once you have classified the item, the next step is to determine whether you need an export license based on the "reasons for control" of the item and the country of ultimate destination.

You begin this process by comparing the ECCN with the Commerce Country Chart (Supplement No. 1 to Part 738). The ECCNs and the Commerce Country Chart, taken together, define the items subject to export controls based solely on the technical parameters of the item and the country of ultimate destination.

How do I screen my customer?

Certain individuals and organizations are prohibited from receiving U.S. exports. Others may only receive goods if the transaction has been licensed, even for items that do not normally require a license based on the ECCN and country or based on an EAR99 designation. There are various lists that may be relevant to your export or reexport transaction such as the Denied Persons List and Entity List. These lists are available on the BIS website at http://www.bis.doc.gov/complianceandenforcement/liststocheck.htm.

In addition, BIS has guidance on its website on knowing your customer and potential red flags in a transaction at http://www.bis.doc.gov/complianceandenforcement/knowyourcustomerguidance.htm.

What happens if I don't get the proper export authorization?

Responsibility for export compliance rests with the exporter. Administrative and criminal penalties exist for violations of U.S. export law. The maximum financial penalty per administrative violation is $50,000.

I'm a small businessperson. Does everyone have to follow these regulations?

Yes. Keep in mind that most low-level or dual-use products will have the designation EAR99 and be eligible for shipment under the NLR designation. Products with this designation will only require a license to certain prohibited destinations, end-users, or end-uses.

If I do have to apply for an export license, can I do it online?

Yes. The fastest way to get an export license is to use the Internet based electronic licensing system, SNAP-R. You must first obtain a PIN prior to submitting an electronic license application or classification request. For further information regarding SNAP-R and PINs, visit the BIS website at www.bis.doc.gov/snap/pinsnapr.htm.

The alternative is the paper application, the BIS-748P multipurpose application which is available from BIS. You must use the original form which must be typed. You can order the form by calling (202) 482-3332.

Is there a fee associated with submitting a license application to BIS?

No. There is no fee associated with submitting a license application or a commodity classification request to BIS.

Assistance is available. The Office of Exporter Services has counselors available from our Washington DC headquarters at (202) 482-4811 and our Western Regional Office in California at (949) 660-0144 or (408) 291-4212.

Is there information available on the Internet?

Yes, additional export control information and related resources are available online at www.bis.doc.gov.

USEFUL TERMS:

Automated Export System (AES) record

The electronic equivalent of a Shipper's Export Declaration.

Bureau of Industry and Security (BIS)

An agency of the Department of Commerce that is responsible for administering and enforcing export controls on "dual-use" items. BIS administers the Export Administration Act (EAA) by developing export control policies, issuing export licenses, prosecuting violators, and implementing the EAA's antiboycott provisions. BIS also enhances the defense industrial base and assists U.S. defense firms. The website is: www.bis.doc.gov

Commerce Control List (CCL)

A list of items subject to Bureau of Industry and Security export license requirements based on their identity. The Commerce Control List (CCL) is found in Supplement 1 to Part 774 of the Export Administration Regulations (EAR). [NOTE: EAR99 items are not listed on the CCL.]

Directorate of Defense Trade Controls (DDTC)

Department of State agency which licenses defense services and defense (munitions) articles. The website is: www.pmddtc.state.gov

Export Control Classification Number (ECCN)

Individual categories of items on the Commerce Control List (CCL) are identified by an Export Control Classification Number.

Export Administration Regulations (EAR)

Regulations set forth in parts 730-774, of Title 15 of the Code of Federal Regulations(CFR), and issued by the Department of Commerce to implement the Export Administration Act (EAA) and other statutory requirements. The EAR is amended by rules published in the Federal Register.

No License Required (NLR)

A symbol entered on the Shipper's Export Declaration (SED), certifying that no BIS export license is required.

Office of Foreign Assets Control (OFAC)

Department of the Treasury office that licenses exports to and other transactions with certain embargoed countries and sanctioned individuals and organizations. The website is: www.treas.gov/ofac

Shipper's Export Declaration (SED)

A document used by the Bureau of Census for compiling U.S. trade statistics and by the Bureau of Industry and Security (BIS) in administering and enforcing the Export Administration Regulations.

Simplified Network Application Process Redesign (SNAP-R)

A method of submitting applications over the Internet with a web browser. To use SNAP-R, one must first apply to BIS for the issuance of a Personal Identification Number (PIN) and Company Identification Number (CIN). The website is: www.bis.doc.gov/snap/pinsnapr.htm

OTHER PUBLICATIONS AVAILABLE FROM BIS:

- Introduction to the Commerce Department's Export Controls
- Guidance on the Commerce Department's Reexport Controls
- How to Request an Export Control Classification Number (ECCN)
- Bureau of Industry and Security

NOTES:

NOTES:

NOTES:

NOTES:

NOTES:

NOTES:

NOTES:

NOTES:

NOTES:

NOTES:

NOTES:

NOTES:

Made in United States
Orlando, FL
07 November 2026

90337525R00015